Neuroscience Notebook

Dan Eitreim
Dan@TheProSays.com
Published by: On Target Publishing
Published in the United States of America

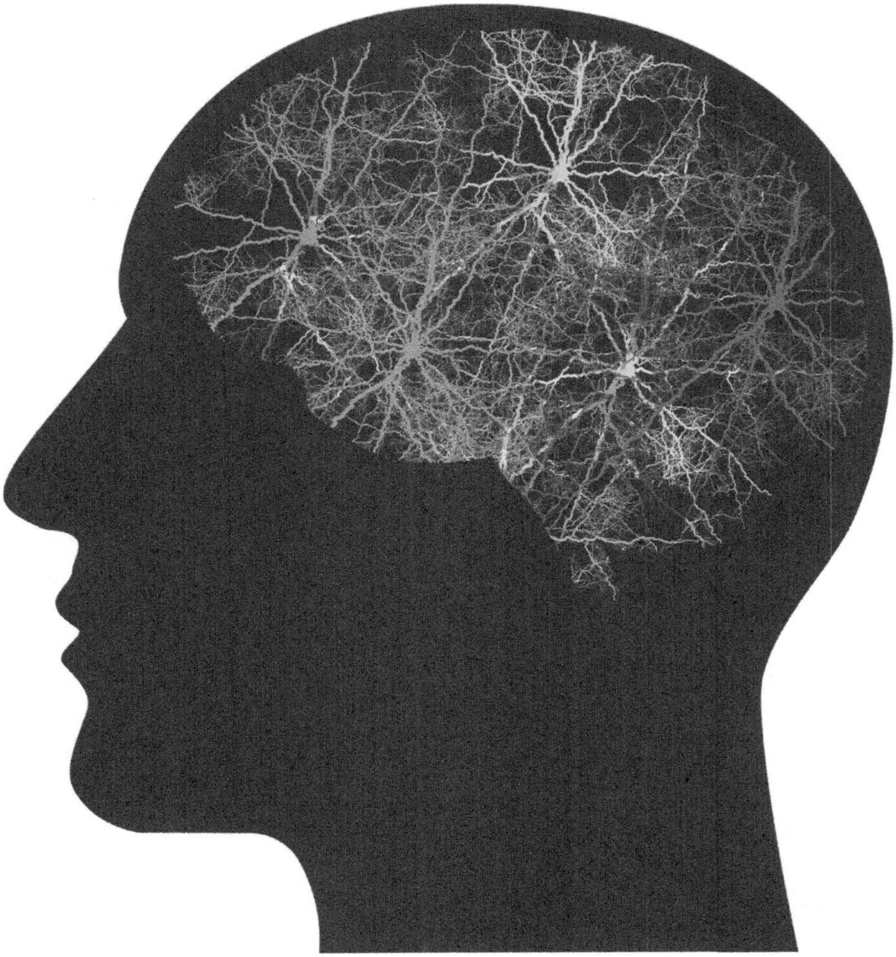

This Notebook Belongs To:

____

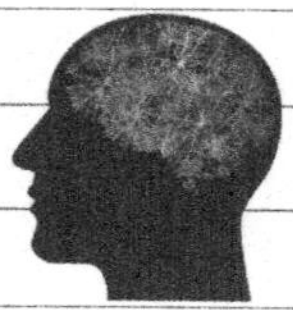

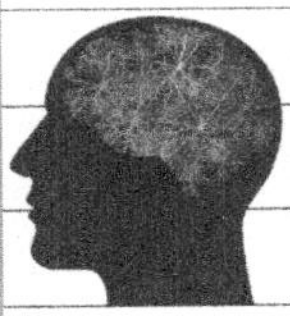

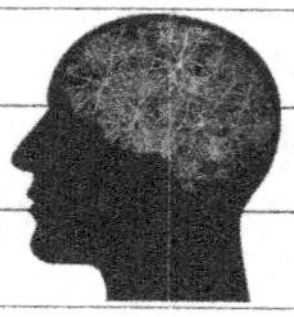

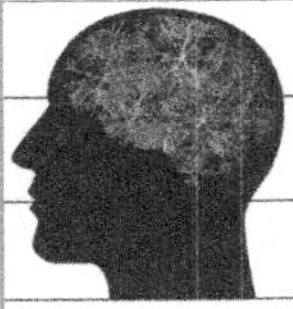

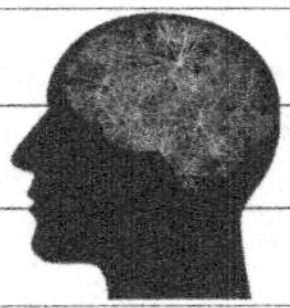

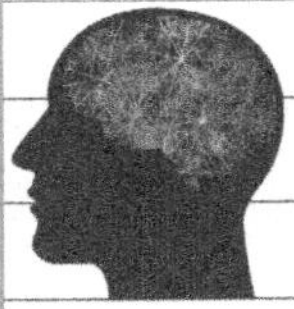

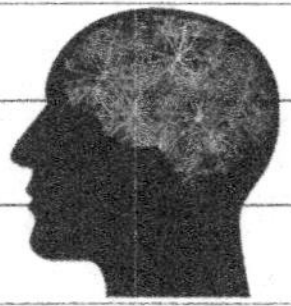

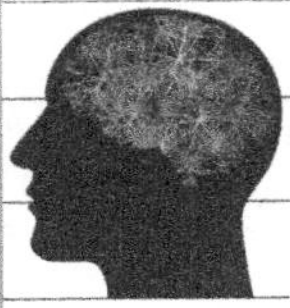

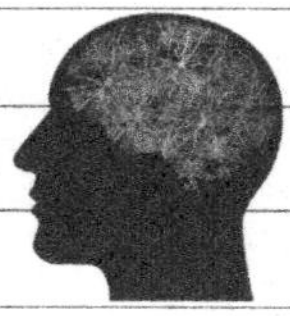

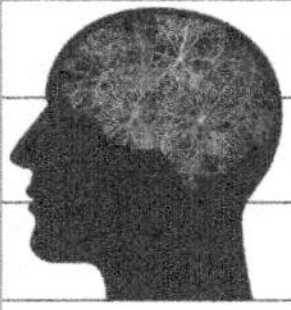

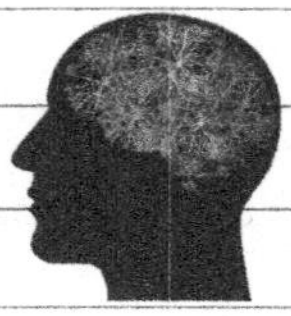

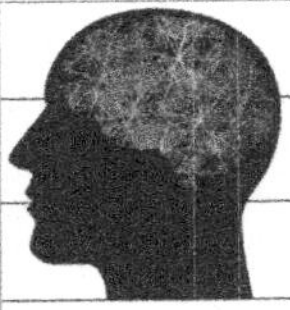

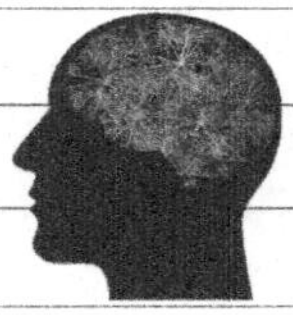

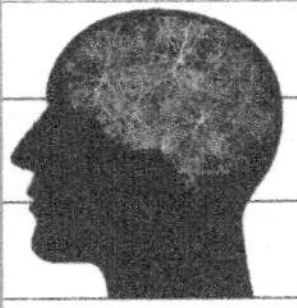

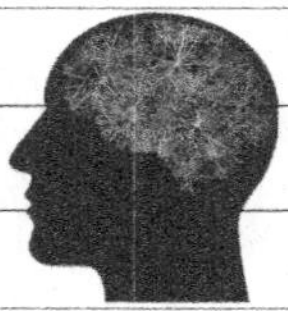

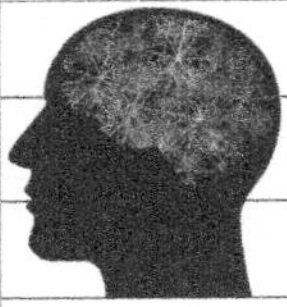

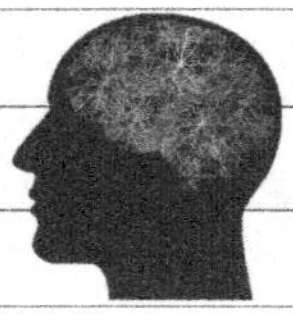

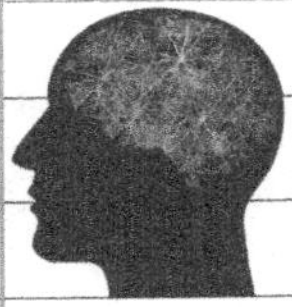

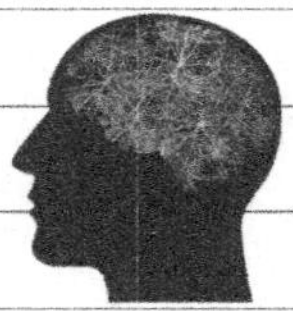

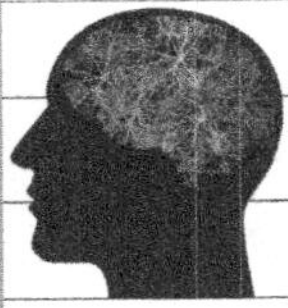

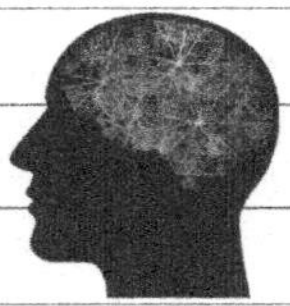

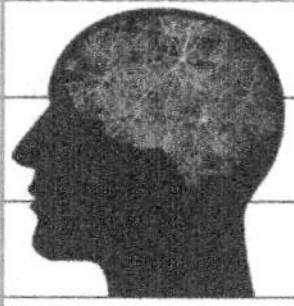

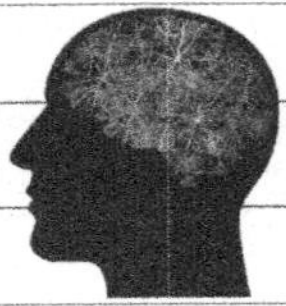

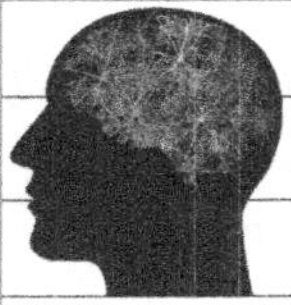

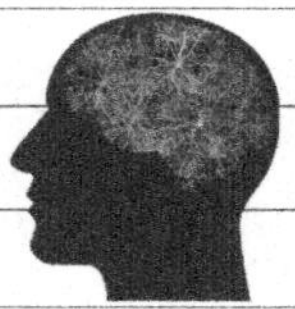

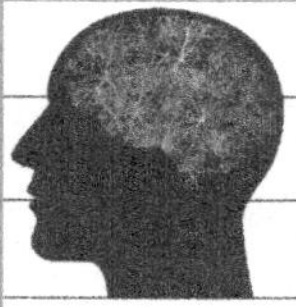

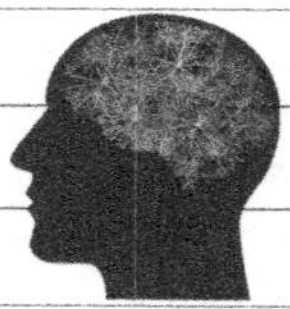

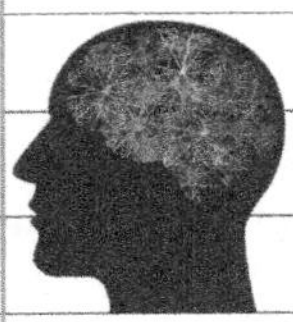

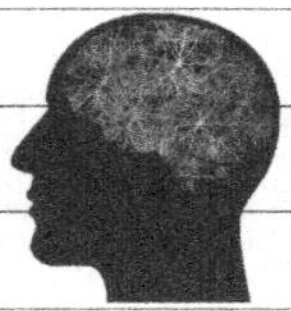

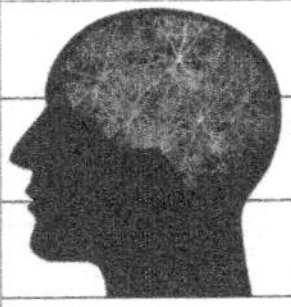

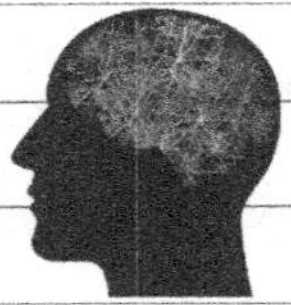

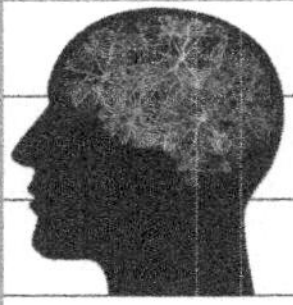

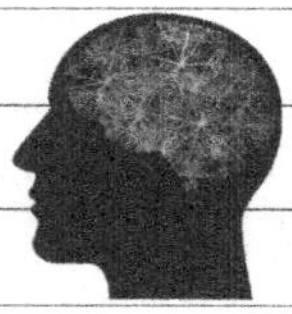

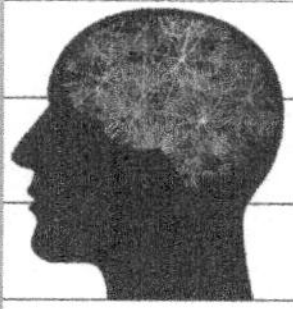

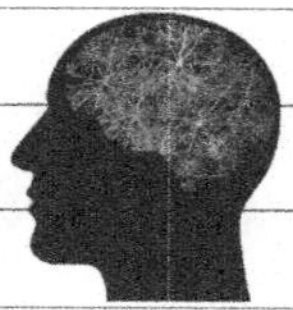

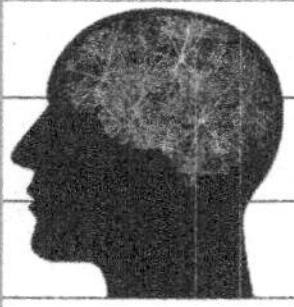

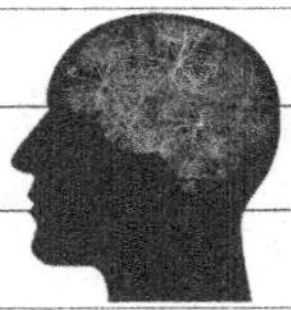

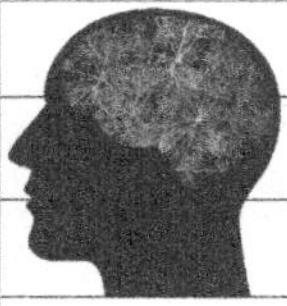

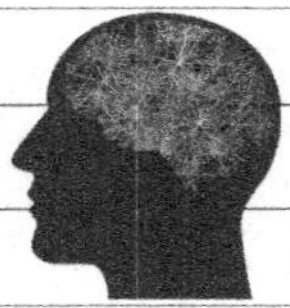

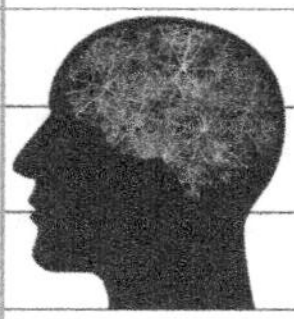

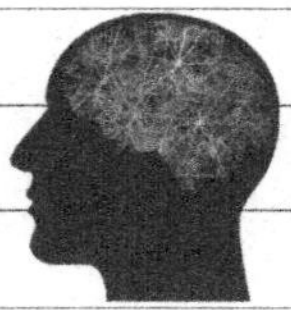

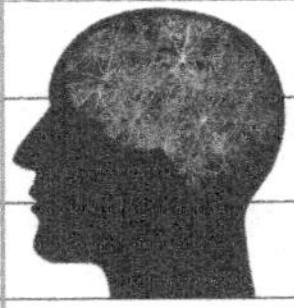

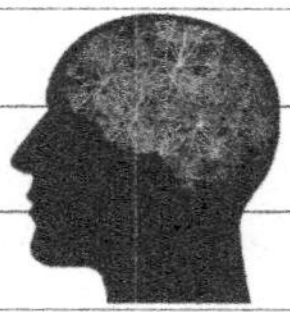

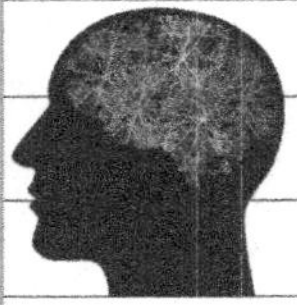

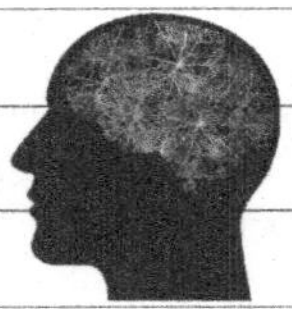

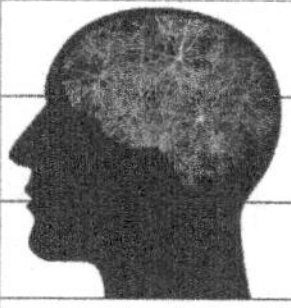

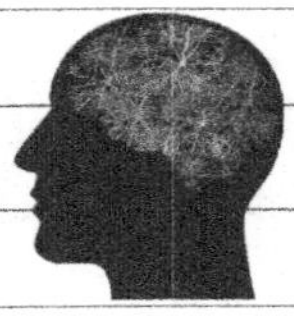

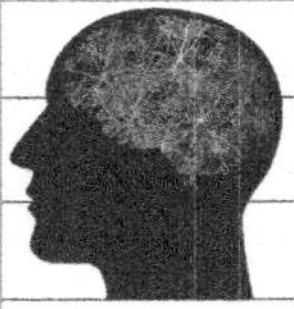

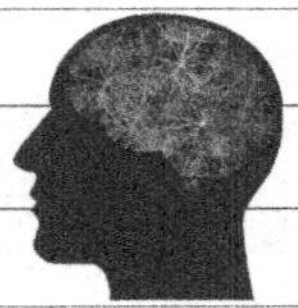

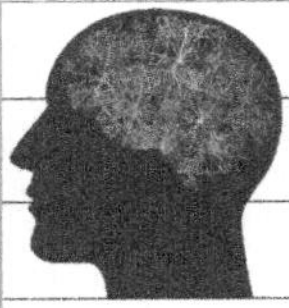

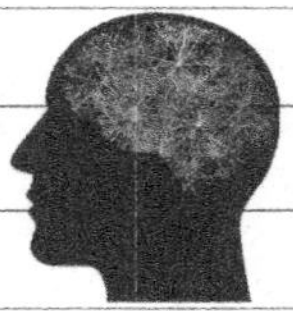

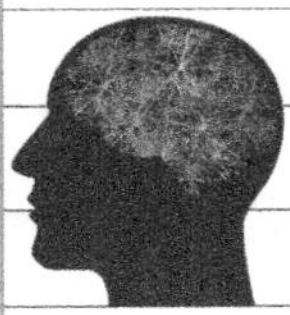

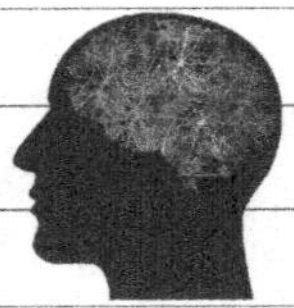

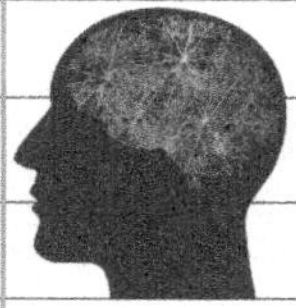

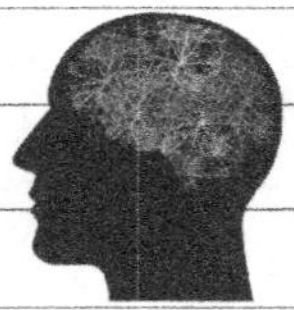

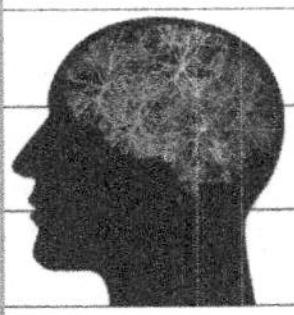

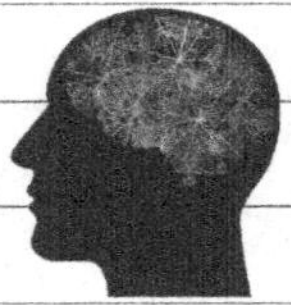

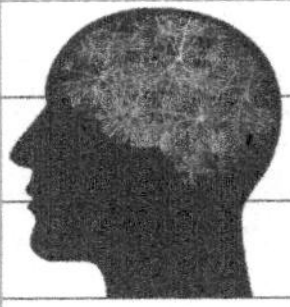

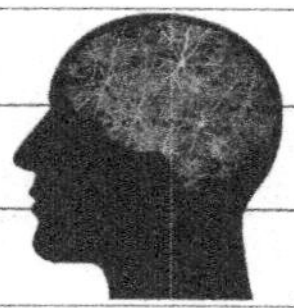

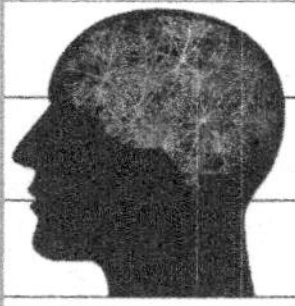

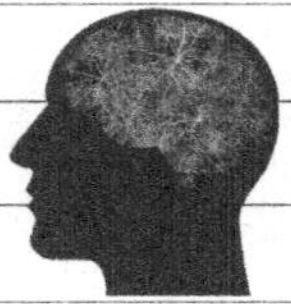

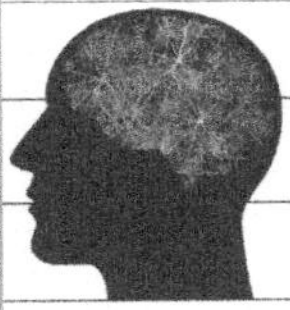

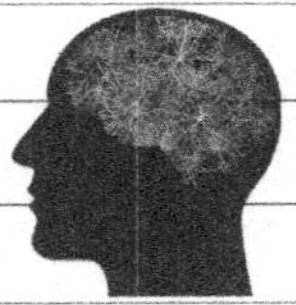

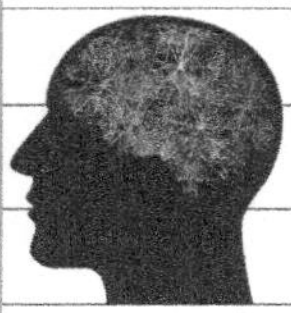

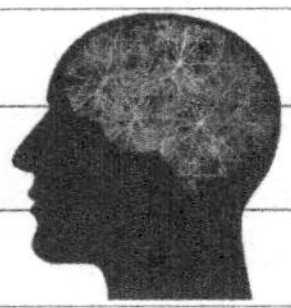

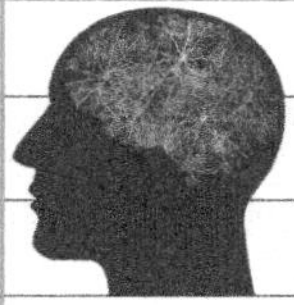

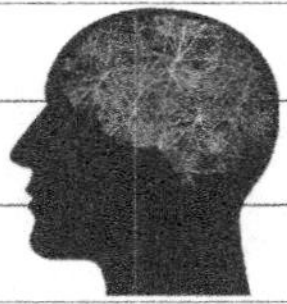

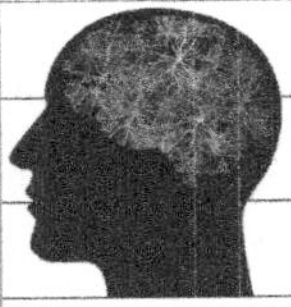

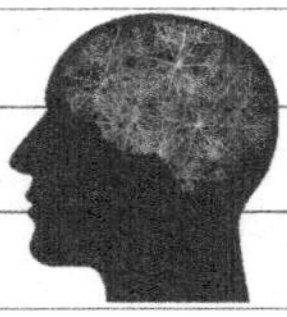

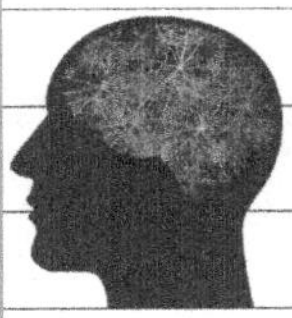

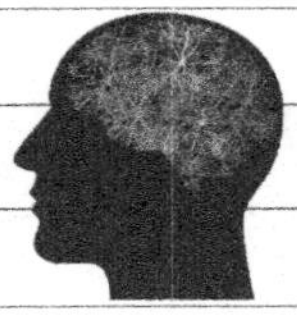

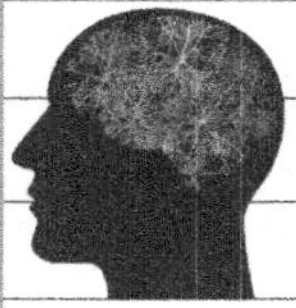

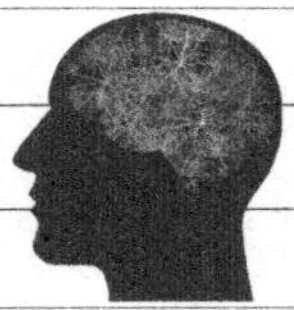

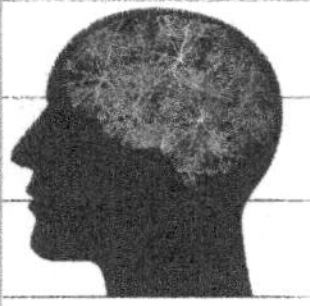

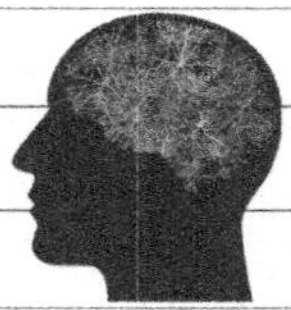

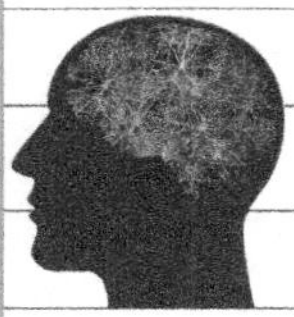

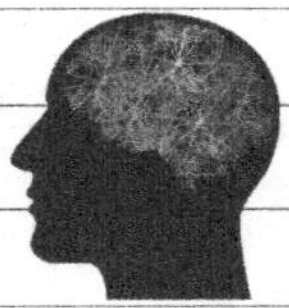

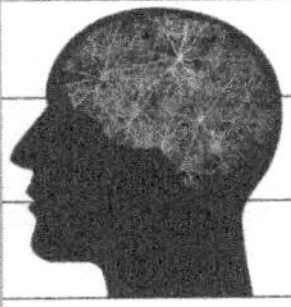

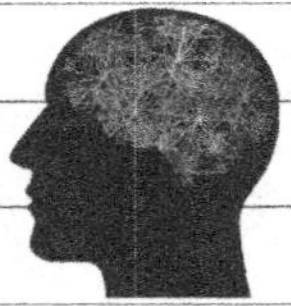

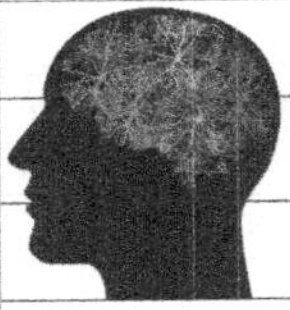

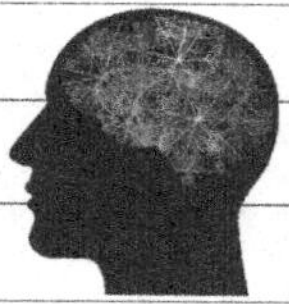

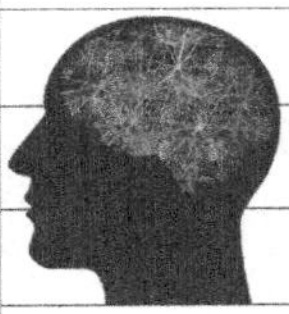

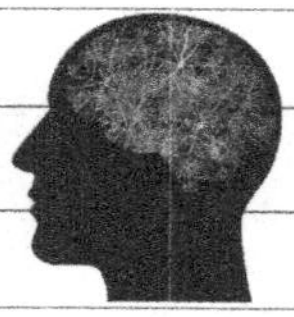

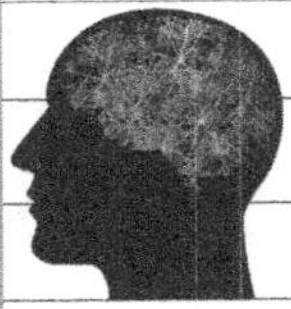

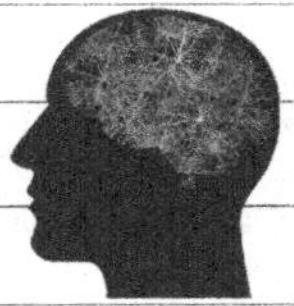

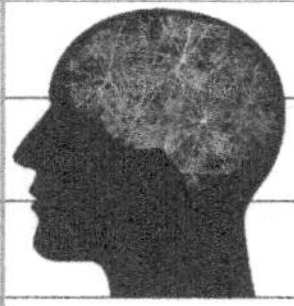

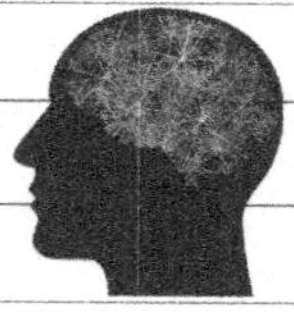

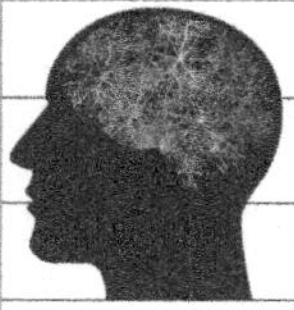

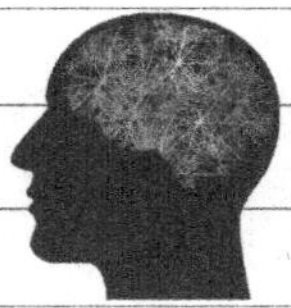

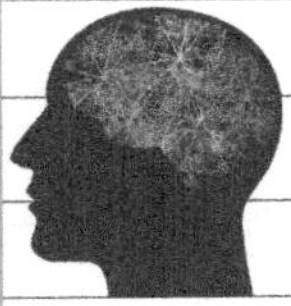

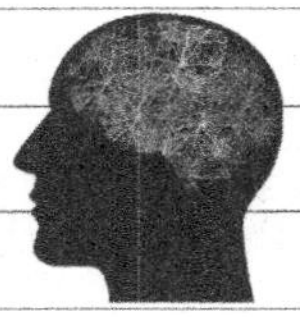

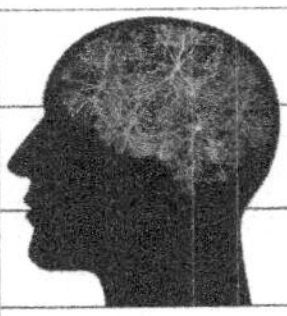

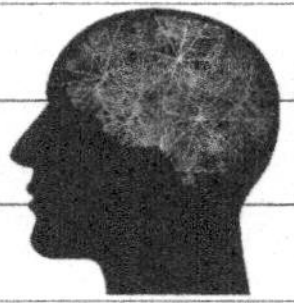

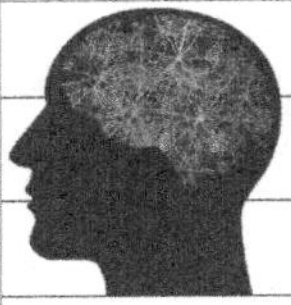

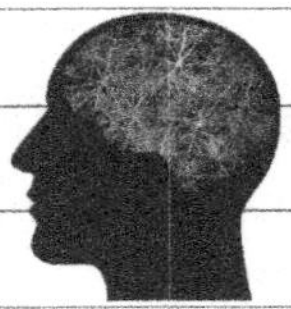

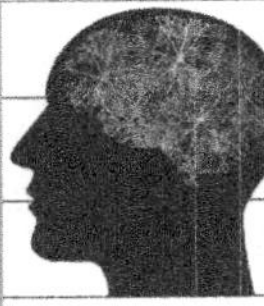

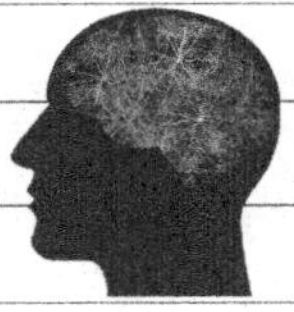

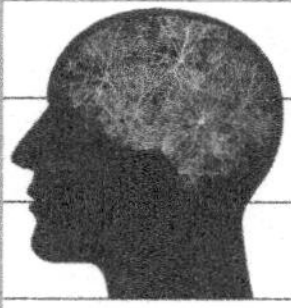

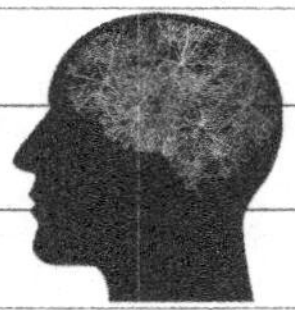

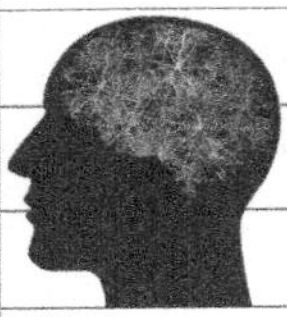

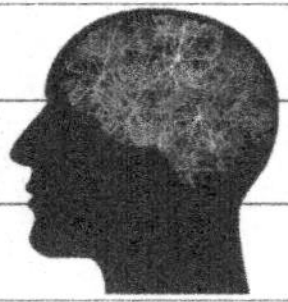

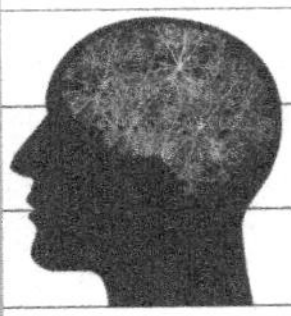

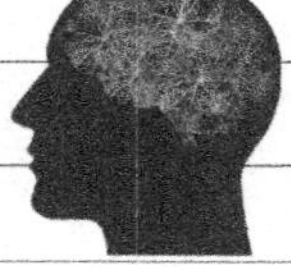

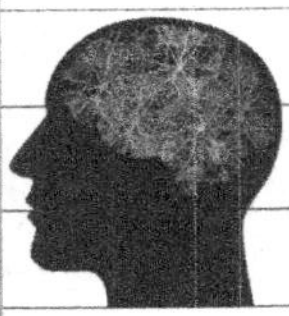

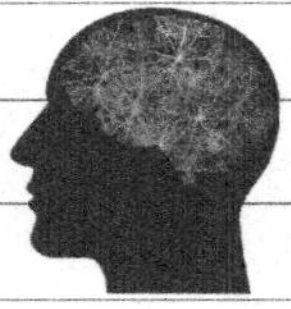

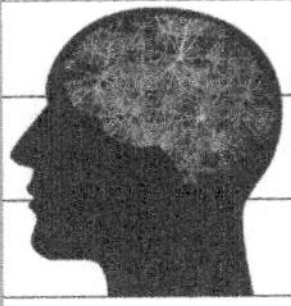

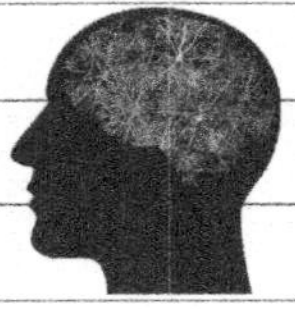

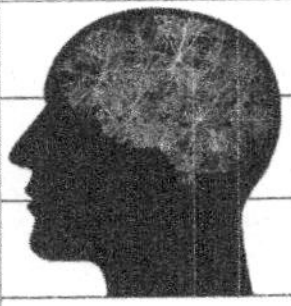

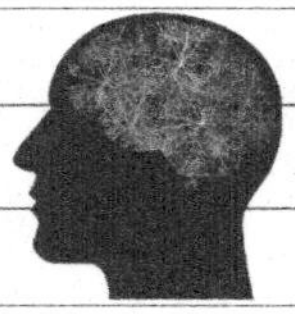

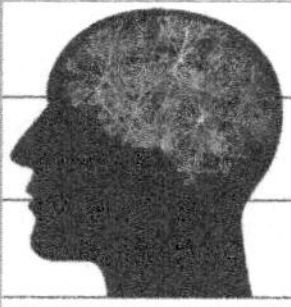

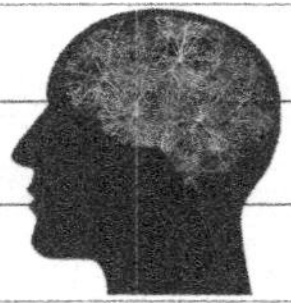

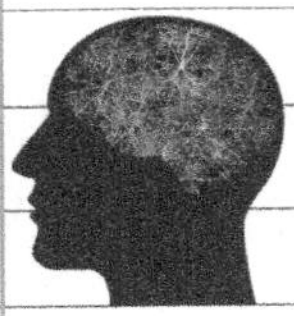

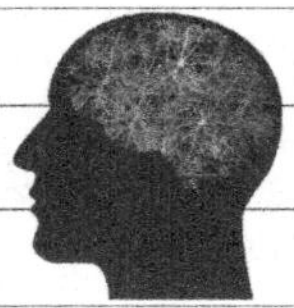

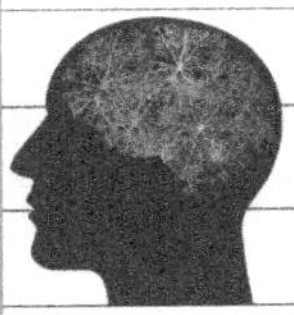

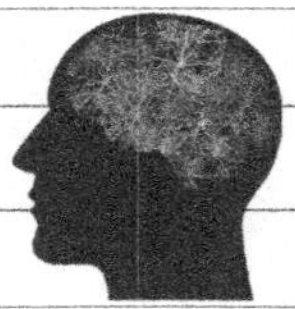

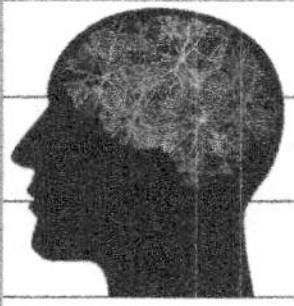

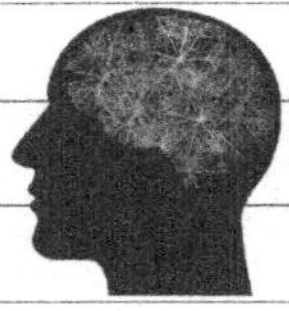

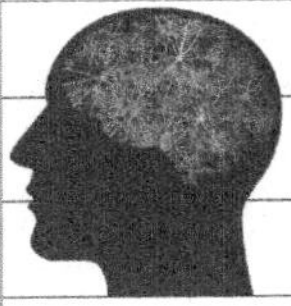

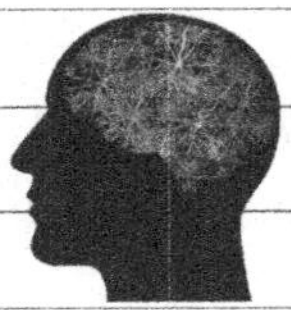

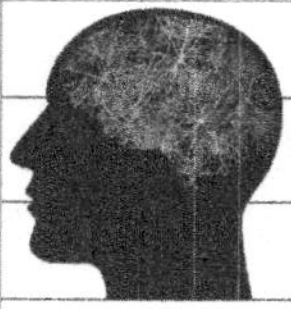

Made in the USA
Middletown, DE
20 October 2022

13144376R00071